BIGFOOT LIVES FOREVER...

IN IDAHO

BY BECKY COOK

COPYRIGHT

Copyright © 2019 Becky Cook
All rights reserved.

All rights reserved. No part of this book may be reproduced, scanned, or distributed in any printed or electronic form without permission. Please do not participate or encourage piracy of copyrighted materials in violation of the author's rights. Purchase only authorized editions.

Cover Art:
Copyright © Brandon Tennant

Every effort has been made to ensure that the stories contained in this book are truthful and honest.

DEDICATION

For my kids and my grandkids -
Thanks for always encouraging me to write down all of my stories and tell them frequently.

"Twenty years from now you will be more disappointed by the things that you didn't do than by the ones you did do. So throw off the bowlines. Sail away from the safe harbor. Catch the trade winds in your sails. Explore. Dream. Discover."
– Mark Twain

CONTENTS

ACKNOWLEDGEMENTS
FORWARD..1
CHAPTER 1: SOUTH CENTRAL...........................1
CHAPTER 2: ARBON VALLEY..........................12
CHAPTER 3: BANNOCK CREEK......................17
CHAPTER 4: BERN..22
CHAPTER 5: BLACKFOOT DAM.......................26
CHAPTER 6: COEUR D'ALENE........................28
CHAPTER 7: DEMPSEY CREEK.......................35
CHAPTER 8: FERRY BUTTE.............................40
CHAPTER 9: FRANK CHURCH WILDERNESS...42
CHAPTER 10: IDAHO CANADA BORDER........45
CHAPTER 11: ISLAND PARK...........................48
CHAPTER 12: MALAD.......................................50
CHAPTER 13: MOUNT PUTNUM.....................55
CHAPTER 14: OVID..60
CHAPTER 15: PALISADES................................63
CHAPTER 16: POCATELLO..............................68
CHAPTER 17: SAWTOOTH MOUNTAINS.........72
CHAPTER 18: STRAWBERRY SPRINGS /MANTUA..77
CHAPTER 19: WILLOW CREEK.......................80

THANKS

With Thanks to:
Kim Auten for amazing counsel and for keeping my head on straight. Gyda Stimpson for being a listening ear for all of my ideas, Linda Andrus for being a wonderful editor, and Scott Armstrong for being the special brand of crazy that I love.

A Special **THANKS** to Brandon Tennant for doing an amazing job (As always!) on the cover of this book.

BIGFOOT LIVES FOREVER...

IN IDAHO

FORWARD

My purpose in writing this book isn't to prove or disprove the existence of Bigfoot, but rather to offer graphic evidence in the form of eyewitness stories that they exist here in Idaho for those who choose to believe. Many folks I talk to have said that the whole Bigfoot phenomenon is a hoax; there can't possibly be a Bigfoot as there isn't any tangible evidence. I think otherwise; there have been visual sightings by reputable, honest, sober folks, and there is corroborating evidence – footprints, handprints, vocal prints, videos, pictures, and hair samples.

For those who have had their own sightings – they don't need more proof; they already know what they saw. It is incredibly amazing to have validation of an otherwise unusual experience, and that is what this book offers.

For those who want to know more – there are websites devoted to the compilation of data – footprints, pictures, sounds and hair samples. This book just offers additional information.

For those of you who choose not to believe – what are you doing reading this book? If we all are crazy, then you just joined the ranks.

Welcome to the club of Bigfoot Believers!

Chapter 1

SOUTH CENTRAL IDAHO

In each of my previous books I have included a story that is from my own experience with the Bigfoot. It's always fun for me to look back at these experiences as I learn so much from each of them. So for this book I am including a few different stories.

For those of you who are reading the Bigfoot Lives book for the first time I need to explain that I am one of the 100 tallest women in the United States and am in the top 200 for the world. As such, I have big feet that are almost exactly 12 inches long. I wear a size 14 women's shoe and somewhere between men's 12-13. I don't often run into people with feet as long as mine but when I do, we are usually talking about where to find clothes or shoes. I found my first Bigfoot print when I was 15 and have seen the Bigfoot personally several times since as have several of my children.

Back in 2012 I had an eye-opening experience where all doubt about the Bigfoot was blown away. At the time I was recovering from a very traumatic marriage, had a bad case of PTSD, and basically lived in fear of anything changing in my life. One night I was awakened and immediately felt there was someone or something outside my home. It was unnerving and my heart was racing. I peeked out the back window and didn't see anything, but I could smell two distinct smells, both of which were unusual. Smells are difficult to stick in a particular category, but one was musty and had a faint garbage smell to it, the other was sweeter and yet still slightly musty. They were both pretty strong smells, but completely dissipated very in short order. I thought for just a few moments they were skunk smells but dismissed that when the smell was gone so completely, so quickly. Skunk odor permeates everything and if you smell it up close and personal you never forget it, besides it tends to give you a slight headache that lasts a few days. So I ruled out skunk but still had no idea what was out there.

A few days later one of my daughters told me that some men came past her window in the middle of the night. Okay, that is creepy no matter how you look at it. But it was even more unusual because we lived in a tiny, small, very rural town where the sidewalks roll up about nine each night. She said she heard the voices directly under her window about three that morning. I asked her what the men were talking about and she didn't know. She said she could hear them clearly, but they didn't seem to be speaking any familiar language. They both had mighty deep voices though; that is why she thought they were men.

That experience was disturbing enough that I called our landlord in Kansas and asked if he had any family who would be stopping by to remove anything from the family storage on the property. He said no one was in the area or even close enough to stop by. Odd…

The next time we watered our lawn my son came in and got me, telling me he had to show me something "really cool." I followed him outside to an area of the lawn where we had flood irrigated just the day before. There in the drying mud was a huge footprint with the heel sunk in about two inches deeper than the toes. I am normally in flip flops, even in the winter, so I pulled off my shoes and put my foot next to this print, and it was a good four and a half inches longer than mine! Keep in mind that my feet are exactly 12 inches long, and these prints that we found easily dwarfed my feet!

Bigfoot Lives Forever…In Idaho

It was a few days later that I put everything together and honestly, even now I am not sure how that happened. I think I had a dream that included the Bigfoot, I don't really know. Anyway, all of a sudden I remembered seeing them from time to time. I remembered some astounding experiences where I felt that they were watching me and keeping track of me – not in an eerie way, but possibly more out of curiosity than anything else.

So I started thinking about the Bigfoot and how to make purposeful contact with them. I don't know how other people make decisions like this but I said a prayer and asked my Heavenly Father. My thought was that God knows everything, and if He created these beings. He definitely knows how to talk to them. After the prayer I felt a strong impression to take some apples out to an area where they were likely to be found by the Bigfoot. I thought I might do some wood knocking and at least let them know I was leaving a treat (the apples) so they didn't sit there and just rot.

So that is what I did. I bought some apples as this was late February and I didn't have any decent apples in my food storage. One evening I drove out to my favorite spot in our little valley where the trees grew tall and beautiful, where a settler had built a homestead and planted roses, but for whatever reason he had left the area and the home had deteriorated back to rock and concrete. It had a pole fence around it and a huge rock where I would go often to just sit and think. I love this

area, still do, and I figured if I was a Bigfoot it was somewhere I would hang out. So I left five apples on the fence in several spots. I arranged them by size – biggest to littlest, and then did some knocking to let the
Bigfoot know they were there, and then I left.

I came back about a week later and some of the apples had been moved. Some of them were missing a bite or two out of them. I figured that was a good start, so I left a few more apples – this time in a cool pattern. I came back after another week and most of them were also gone. The next time I left five more apples – one of which was HUGE. I put it up near the center of the other apples and this time I stayed and did some thinking and singing because when I am outside in the woods I usually am singing. Eventually I went back home and came back another week later. This time most of the apples had been eaten, but the huge one had one bite eaten out of it and then put back. That one bite was enormous though, taking in almost half of this great big apple, far bigger than a human would have left. It made me laugh – at least they liked the apples!

It was about this time that the work I was doing for a living ended with a strong element of finality. That night as I said my prayers I asked Heavenly Father what He would have me do to support my family. I am the sole provider for my family and that responsibility weighed heavily on my mind. In the middle of the night I was shown clearly the cover of a book with my name on it as author – *Bigfoot Lives in Idaho*. I woke up the next morning feeling comforted and excited to be able to collect more stories about the Bigfoot.

I have actually been collecting Bigfoot stories since I was a small child. During the time I was growing up on the Fort Hall Indian Reservation, I heard all sorts of stories which I wrote down and then later in my life when I went to school near Blackfoot, Idaho, I heard even more stories. Once I started earnestly looking I found a lot more and have since learned a lot.

Which brings me to this point in my story, because you see, there is more.

Once I knew the Bigfoot were in the same valley as I was I determined I would like to see them closer than I had previously. I organized a Bigfoot conference that summer and invited several speakers to present. When I initially asked them to speak, I didn't know much about them – Thom Cantrall from Washington state, Arla Williams from Oklahoma, and Scott Nelson from Kansas. I had this ulterior motive though – after their speaking engagement, I would have them hang out at my place and maybe, possibly, we could see the Bigfoot.

In the world of Bigfoot enthusiasts there are basically two camps or styles of thinking – scientific, and spiritualistic. The scientific people want proof, and honestly I have seen some seriously cutthroat behavior from them. The point is this – THEY want the proof for themselves, but if someone else finds proof they will find some way to downplay or degrade that finding. In the spiritualistic world there are many who have seen the Bigfoot and are satisfied that they know what they are seeing – they don't necessarily need to prove it to anyone. There are still those who doubt and who aren't pleasant about their comments, but for the most part they know what they know and don't worry what others think about it. I fall into that last category because I KNOW what I have seen, I KNOW what I have heard, and I don't feel the need to validate those experiences for anyone. It isn't likely anyone will make a movie out of my life, but at the end of it I will know that I have lived the best life I can in honesty and with integrity.

When those speakers came into Idaho for the Bigfoot convention, I invited them to my place for a smaller get-together. We sat and talked and the next day we went up to the hills and looked at the area where I had seen and experienced the Bigfoot. We took along a picnic lunch and lawn chairs, and as lunchtime rolled around we sat next to a gurgling spring and talked. Arla had her drum with her, and after a while she played for us and sang, and of course we sang with her. Before too long she glanced back at the six of us sitting in the lawn chairs and said, "There's one!" and before too long said, "There's another and one more!" she pointed up the mountain from where we sat, and there they were, three Bigfoot, literally 25 feet away in plain sight, in the middle of the day in nearly full daylight.

It has been a few years but I can still tell you what they looked like and how they moved. The one closest to Arla was huge, and we couldn't see his entire body, but the part we could see was enormous. The other two were younger and were still growing; frankly they looked like skinny teenagers with protruding knees and elbows. The big one was massive though.

They were only there a few minutes and then they were gone. What a memorable experience! I am grateful I was there that day.

Later that summer I took a honey dew melon over to share with the Bigfoot. I was back in the favorite spot over by the old homestead, and this time I was hoping to find a safe place to put a melon where it wouldn't roll if I set it down. As I was looking around, I noticed that one of the older tree limbs had fallen off up in one of the trees near the homestead. It left a hole in the tree trunk like you would picture a squirrel using for nuts and I thought that would be perfect for leaving the honeydew. So I walked over to the tree and stood with my belly right next to the trunk. Stretching my arms up as far as I could go, I rocked that honeydew into that hole in the tree.

Bigfoot Lives Forever…In Idaho

As I stepped back, I thought how picturesque that honey dew looked – like a green eyeball on the trunk nearly eight feet above the place where I was standing. I left again and came back in a week to see what I could see. The honeydew was gone and there was no trace of the rind or anything like that as there would have been with a squirrel or another animal eating it. Down below the tree in the deep dust of late autumn were two ENORMOUS footprints! Each of them was easily 20 inches long and about ten inches wide. I had never seen footprints that large prior to this or even since this time. I can only surmise that the footprints I found were from the big guy that I saw earlier that summer. I didn't have a camera or a cell phone with me but once again, I knew what I saw. What an amazing day!

A few months went by and my family all came home for the holidays. Since it was a mild day, we all went for a walk around the park. Usually in the middle of the winter there is plenty of snow, and it isn't likely anyone would even suggest a walk outside. This particular day however was beautiful and relatively balmy – a wonderful day to go for a walk.

We set out around the park and were halfway when I noticed a footprint, frozen in the icy grass. I pointed it out to one of my daughters, and we paused long enough for me to remove my flip flops (Yes, I wear flip flops in the middle of winter) as I carefully measured it next to the frozen one. The footprint was about four inches longer, frozen in the icy grass as clear as though it were cast in plaster. I glanced up, across the park and could easily see seven or eight footprints cutting across the park, as clear as could be. They began near where the park passed near the main road and appeared as though this particular Bigfoot had been traveling close to the roadway when a car came towards him, and he cut across the park. By the time he had reached the point where I saw the footprints, he was running, as the footprints were spread out and pushed deep enough into the ground that it would indicate the higher rate of speed. Absolutely amazing!

My daughter and I both just giggled at our inside joke – how many people might have seen a Bigfoot that particular day if they had been aware they were that close? I think it is truly a matter of being aware of our circumstances and keeping our minds open to the fact that we share this land with other beings. That, and a little luck, and I think others will see the Bigfoot.

I know many of you reading this will want to take these stories with a grain of salt and that is okay. That is your prerogative. Just be aware that someday when you tell about your own experiences, you might also face disbelief from listeners. Always remember that you know what you know and you don't have to prove anything. Good luck!

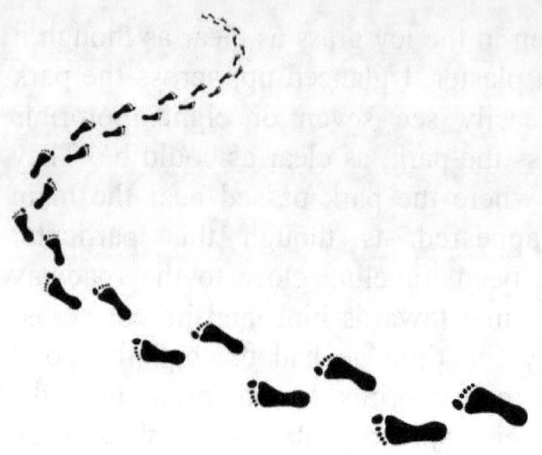

CHAPTER 2

Hunting. It's one of those enjoyable male (and female) bonding moments when you are out in nature on the side of a mountain. Sometimes you freeze, sometimes you nearly melt, but most people continue to do it year after year in spite of the cold or discomfort.
Eric Badger is one of those men. The meat is a great incentive, but he also just enjoys being in the mountains with his friends, at least until something weird happens.

ARBON VALLEY

It was October 2012. Eric Badger had been looking forward to going hunting with his good buddy Mike up in the hills above Arbon Valley. The day finally came. They loaded up the pickup, drove to their favorite camp site, and set up. After eating, they retired to their tent to get some sleep when they heard a crazy weird noise coming from the mountain beside them.

"We could hear something coming down the mountain, making a lot of noise," Badger said. "We couldn't see what it was, but then we weren't too anxious to go outside as it was blowing like crazy, and we were freezing."

They stayed in the tent and prepared to go to bed when they both felt something was in their camp outside. They would hear objects move around, but softly, carefully. They had left their cooking stuff on a table outside, but none of the supplies or even the dirty pots were disturbed.
"I didn't see what was out there. I didn't see anything, but I also know that if it had been a bear or another animal, it would have gone after the dirty pots looking for leftover food."

Badger said there was a rogue dog up there, a Great Pyrenees that wandered around. It wasn't friendly and at times would make an appearance that would startle the unsuspecting. However, he said he was also positive this experience wasn't due to the rogue dog because of what happened next.

"We could sense that whatever was out there was walking around the tent, but the ground was so frozen we couldn't hear anything," he said. "Whatever it was went clear around the tent though and didn't trip on any of the guy wires. It didn't stumble at all. As it passed the doorway of the tent though, it REALLY smelled bad."

He said that it smelled like a big pile of cow crap, like something had died. They still hadn't seen anything, so they didn't know what they were dealing with.

"I wish I had been driving my other pickup with the automatic lights because then I could have just turned on the lights and seen what was out there. We would have at least seen a shadow." he said. "I was going to turn on my big flashlight, but my buddy Mike said not to – he was afraid I would tick it off, whatever it was."

They didn't have their guns with them in the tent either – they were all in the pickup. He was kicking himself for that. About that time something felt around the outside of the tent and then came down the side of the tent to where Mike was lying on his bed, making contact with his leg through the wall of the tent. It freaked him out and it was then that they bailed completely.

"We just bailed out of the tent and jumped into the pickup and left," Badger said. "We went down and around the mountain and ended up sleeping in our truck on the other side. We just left everything there. We came back the next day and nothing had been disturbed, it was just freaky."

Bigfoot Lives Forever…In Idaho

Badger said he asked Mike what it felt like when the being touched him and Mike said it felt like a human hand – like someone had reached out and put their hand over his leg. That wouldn't be odd or weird except for the fact that there weren't any other humans out there that they knew of. Even if it had been a human, they would have needed some type of light to make their way around the tent without stumbling or tripping on the guy wires of the tent.

"Keep in mind, we never saw what it was," Badger said. "But that isn't bear country and besides a bear would have been making more noise. We would have heard something as it went around the tent or tripped over the wires on the tent. It would have disturbed the pots or gone after food scraps. The ground was completely frozen but if it had been a horse or something like that, we would still have heard the hooves on the ground. We didn't hear anything like that."

Badger said this wasn't the first time that something like this happened in that area to him. Once before he had been camping in roughly the same spot when something came through his camp when he was camped out and hunting.

"Two or three years before this, something was walking around in our camp but we couldn't see who it was," he said. "We got dressed and put our boots on and bailed out to check what was out there. Nothing. We couldn't find anything, but there was definitely something in our camp, we both heard it."

That time they ended up sleeping in their pickup - at least it felt safer than being in the tent on the ground.
Badger said that he spends a lot of time up in the hills whether for hunting or just enjoyment and one other occasion he had something odd happen but it was up by Stanley Basin.

"I was up in the mountains with my friend Barry and we were on horseback," he said. "We both heard someone throwing rocks off the hill and down the mountain."

He said that often bears will roll stones over while looking for grubs and they might roll down a hill, but these were definitely thrown.

"I'm not going to say that what we had in our camp was a Bigfoot, I don't know what it was," Badger said. "I just have no other explanation that fits it."

Chapter 3

Members of the Shoshone Bannock Indian tribe often have sightings of the Bigfoot but they don't always mention anything to anyone outside the tribe. I heard this story while talking to another tribal member and ended up tracking the story down so I could retell it here.

BANNOCK CREEK

Kathiana Eagle was with a family group when they took her 1999 Quad cab pickup hunting up near St Anthony one day. They shot three elk and brought them home and then spent the next few hours gutting them out and hanging them to bleed. They had mostly cleaned out the pickup, but there was still some blood in the back of the truck that they hadn't washed out yet.

The next day they were out Bannock Creek with the same truck, and they were gone all day, coming home late at night with three dogs sitting in the back seat. All of a sudden the truck started losing power and the RPMs went way up, and about that time she felt something hit the back of the truck.

"The dogs jumped over the seat onto the floor of the passenger side of the truck," she said. "It was pitch black outside, and I couldn't see him, but I knew he (the Bigfoot) was there. He stayed there about a quarter to a half mile until we hit the Arbon Valley Road, and then I felt him jump out of the back."

Immediately the RPMs went back down and the dogs relaxed.
"I was freaking out! It was pretty scary!" Eagle said. "I flew down the road to Bannock Creek, and when I stopped later we took a good look at the back of the truck."

She said she and her husband found clear hand and foot prints in the truck and could clearly see where the Bigfoot had licked the blood out of the grooves in the truck bed.

"I took pictures of all of it but the pictures never developed," she said. "That seems to happen with pictures of the Bigfoot."

Eagle's job involves delivering newspapers in the middle of the night to drop location from where they will be delivered to customers the following day. On one particular day she was coming home as the sun was coming up when she clearly saw a Bigfoot sitting about a hundred feet off the road.

"He had his foot crossed over his other leg and he was picking something out of his foot – I can still see it in my mind's eye as he spread his toes apart," she said. "He was dark brown with a reddish tinge, and he had his head tilted at an angle so I couldn't see his face. He was just pulling something out of his foot like we might do when we are walking around outside barefoot."

She said she often sees the Bigfoot on the way home in the early morning light. Three different times she has seen him on Siphon road in the dip just before it meets up with the freeway.

"The last time I saw him there, he was standing along the side of the road," she said. "There was a bush right there and he was apparently trying to blend in, but I got a good look at him as he stepped over the guardrail heading north. It was just a short glimpse, but he looked pretty beat up."

These footprints were sent to me from a friend who found them near the side of a roadway. He said they were about four and a half feet apart and obviously barefoot, which is the first aspect he noticed. Who goes barefoot in the middle of nowhere in the snow?

CHAPTER 4

As is often the case, people who have seen the Bigfoot know other people who have also seen the Bigfoot. Such is the case with this story. I was told about Ron Harper's experience through Michelle Higley, whom I heard about through Dave Higley, her husband, (Story found in Bigfoot Lives in Idaho, Book One) Ron's brother David and good friend Craig Kunz have also seen the Bigfoot down by Bear Lake, and their stories are also included in this book.

Both Harper brothers made the comment that they told their dad, but he didn't seem to become visibly excited about it. It makes me wonder if he knew the Bigfoot were there because he had seen them himself...

BERN

Bigfoot Lives Forever…In Idaho

Ron Harper grew up in the northern most house in Bern, a small community down in the southeast corner of Idaho by Bear Lake. In 1978 he was eight years old and spent a lot of time playing outdoors. On this particular day he was out with a pellet gun, shooting at ground squirrels. It was a warm, clear sunny day in spring.

Directly behind his home was a hill approximately 300 feet higher than the area surrounding the house. There wasn't much grass because it was early in the year and the rest of the area was covered with sagebrush and rocks – it was a pretty remote area.

"I glanced up and distinctly saw a dark brown being, much taller than a man, with very long arms," he said. "He walked up that hill faster than a grown man could run. He was very tall with a huge stride."
Harper said he went inside and told his father what he had seen and his father reassured him that other people had also seen the Bigfoot.

"He didn't get as excited over it as I thought he would."

* * *

David Harper is Ron Harper's brother and they are another family where they both had sightings but separate experiences. Since they both grew up in Bern - a small, rural community to the west and a little north of Bear Lake. It is honestly one of those places where if you blink you can miss it completely, but apparently it is perfect for seeing Bigfoot because several people have had sightings there.

BERN

David Harper had a close encounter with a Bigfoot and it left him with an eerie feeling for several weeks. It happened during the summertime in the mid 70's when he was a youngster of twelve or thirteen years old. The home he grew up in had a circular kitchen with a bar in the middle and the children were running around it in a big circle.

"I just happened to glance up and saw a Bigfoot looking in the window at me," Harper said. "The window was set up in the wall a ways with a curtain on the bottom so it had to be a pretty tall Bigfoot to even see into the room."

He said that it had a thinner face around the eyes than pictures he has seen depicting the Bigfoot. It also had cheeks that looked fuzzy, like those of a black man with black skin underneath like a African man.

"I screamed to my dad "Dad, there is a Bigfoot outside – lets go out and see it!' When we went outside it was gone," Harper said. "I can't remember if we found footprints or not because it would have been standing in the flower bed."

Harper said his father mentioned that other people had seen the Bigfoot in their area around the same time. He also remembers having an eerie feeling for nearly two weeks after that – almost like someone was watching him. Their home was the last house at the end of the road with a mountain range behind them so it would have been a simple feat to watch the family from nearby trees.

He said he has continued to look for Bigfoot when outside anywhere, but he hasn't seen anything remotely close. "Over the years I have told the story to a few people, but they think I am a liar," he said, "I've never seen another one even though I've kept my eyes open for the possibility."

One thing is for sure, it was eerie. "It was pretty spooky! I don't know that I want to see one that close again," he said. "The biggest difference now is that I am not a believer – I am a knower. I know they are out there!"

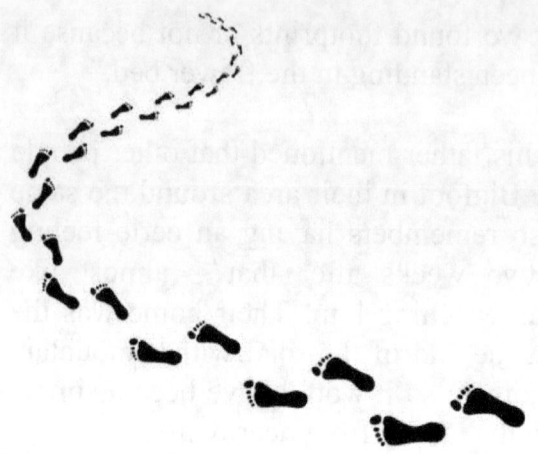

Chapter 5

I was introduced to Delane Crofts through his son who told me a little about the Bigfoot experience his dad had when he was a kid. It sounded interesting, so I tracked Mr. Crofts down to talk with to him.

BLACKFOOT DAM

Delane Crofts lived by the Blackfoot River about six or seven miles below the Blackfoot Dam on a family farm during the 1980's when he was in his early twenties.

On an early spring night his family – mom, dad, sister and he – had just come in from the field after plowing all day. It was getting dark when their dogs just went nuts, barking and growling, and they ran up on the porch where they were overly excited.

"We couldn't see anything but figured it must be a bear or a mountain lion as we had a few of them roaming around," Crofts said. "My dad opened the door and the dogs came tearing in and cowered underneath the kitchen table. A short time later all the cows came running in from their field bellowing and bawling – it was just weird!"

He said that he and his dad jumped in the pickup with their rifles and shined their lights all over but couldn't see anything. Eventually the cows settled down, and everyone went to bed for the night. The next night when they went to the field, they found three sets of big footprints – one set was eighteen inches long, another fifteen inches, the last much smaller.

"We left them undisturbed and later Dr. Meldrum (Bigfoot researcher and Biology professor at Idaho State University) came over and cast them," he said. "They went straight across the freshly plowed field we were in and up into the sagebrush. Just as they started up the hill, the midsized footprints got much deeper and the smaller set disappeared."

He hasn't seen anything like that before or since, but he continues to hope he will see something like that again. It's a memory that Crofts has shared with his children over the years, and he still enjoys hearing Bigfoot stories.

Chapter 6

Tim Flowers has seen a lot of Bigfoot activity near St Joe's National Park when he was a child. Later when he was older, he saw more activity near Lake Coeur d'Alene further downriver. Tim believes this is a long-term relationship, and he is hopeful he can become better acquainted with his Bigfoot neighbors.

LAKE COEUR D'ALENE

"I actually saw the Bigfoot several times in the St Joe's National Forest as a child, and that is what made me go looking as an adult," Tim Flowers said. "The first time I remember seeing anything I was creek fishing and I heard a boulder move. I looked over there and saw the back of a Bigfoot, going into the forest."

It was black and since he was seeing it as a child, he thought it was a giant. "I thought it was very tall – about six or seven feet," he said. "I ran back to our camp and told my parents what I had seen, but they didn't believe me when I said it was Bigfoot. They thought that it was a bear, but I had never seen a bear walk on two legs!"

He said after that he would look for footprints scattered out in the woods with his brothers, and several times they found some.

"We were out elk hunting one year when one went across the road ahead of us," Flowers said. "I saw the back of its head, shoulders and butt going into the woods."
He said his dad saw the same view and said, "What the heck is THAT?"
"I told him that was what I saw and had been seeing since I was a kid, but my dad didn't know what it was at the time."

A few years later he was out fishing once again with his friend, and they were working their way up the creek when they both "limited out" on trout. Just then they spooked something and it took off.

"I saw the back of this one again and it was black," Flowers said. "We heard it jump the creek above us where it was twelve feet wide and we found the footprints on both sides of the creek. It went up the mountain, and we went back to camp where we packed up and took off."

He said back then when he would hear wood knocks, his brother always said it was elk. Flowers didn't believe him then and over the last few years he has formed the opinion that the Bigfoot use the knocking to act like their early warning system, because they would usually put a tree across the road if they are ignored.

That same year he went elk hunting and was coming out of a canyon a mile from their camp. "I was walking down a logging road, headed back to camp, when I heard a scream so loud it went clear through my body – it was like being at a REALLY loud rock concert," he said. "It sounded like a female screaming at the top of her lungs, and I dropped to my knees. I looked everywhere but couldn't see anything."

Flower explained that a few years later he spent eight years in the Marine Corp and when they would fire off rounds the percussion went through his body like a wave. He said the effect of the sound he heard that day in the mountains was quite similar to the one he felt years later in the Marine Corp, but it seemed like it lasted forever.

"I looked around and I had my gun safety off because I thought I was in extreme danger," Flowers said. "I have heard mountain lion but this sound didn't compare to it – it sounded like a woman in distress but I didn't think it was actually a woman."

He said after being in the Marines he took his girlfriend up to his old stomping grounds and went back to the same area in an attempt to locate the being that had made the noise back then. By then the area had been so logged out there was nothing much there. He said they spent the second day just driving his Jeep around and looking at the snow in the hills.

"I found some great big holes in the snow then," he said. "They were off the logging road and I got a bit scared. I took some pictures, but I was just amazed at the size of them. It was then I realized I was actually seeing signs of Bigfoot."

He said he had his gun with him but realized the size of the gun wasn't big enough to do anything to the Bigfoot if he hit it, so he ruled that out. Eventually, he turned around and drove up another logging road where he found a cow moose with is spine and head on one side of the road and the carcass on the other side of the road. No meat had been taken from the carcass, just the innards. It was then he decided to "pack it in" and go home.

He came back to the same area two weeks later and drove to where he saw the footprints. He said he stopped the Jeep and was looking around when a cow elk came right by the Jeep, looking up the hill, completely distracted.

"I got out of the Jeep then with my guns and walked into the woods for a little bit. I heard one long drawn out WHOOOP up on the left and there was another on the right and another down below us," he said. "It was eerie! I emptied my gun into a dirt bank to just scare them away."

A couple of mornings later he left some apples on a stump by the creek. The next morning they were gone, but a big rock had been left on the stump.

"I took a picture of it, but I left because I got the heebie jeebies – the feeling like I got in the Marines when death was going to happen. I showed the picture to a friend later and he said he saw a face in the forest that looked Native American, back in the woods." His friend described it as a cross between Native American and troll with a little bit of logger thrown in for good measure.

Flowers said the last two summers he has looked but didn't see anything, however the forest service had started logging off that area. He actually went up to where the rock was and pushed it off and then took pictures all around that area. He said he returned to where he was staying and looked at the pictures and could see several Bigfoot in the picture. Two weeks later he went to the same place and took his binoculars, and took the previous week's pictures and identified several of the Bigfoot in the nearby area.

"I watched them with my binoculars and took pictures with my cell phone," he said. "They watched me and didn't move for almost ten minutes. I started feeling a little uneasy and realized there was one more somewhere, and he wasn't there where they were. He's bigger and has more hair on his face. I made my way back to the Jeep and they never took their eyes off of me."

He said two weeks later he drove over again, but he was tired from work. Flowers parked his Jeep to sleep for about an hour and was awakened by whistling – one was in back of the Jeep and one was in the front. The one in back was less than a hundred yards behind him, just squatting down. Later he returned and took pictures of the hand and foot prints.

Flowers continues to travel up to the same area often in hopes he will see the Bigfoot and maybe someday have a greater interaction, but for now he has no doubts as to what he is dealing with.

"These things are real!" Flowers said. "I've seen them and I believe in them, but I haven't been hurt by them!"

Looking for Bigfoot is like looking for a needle in a haystack – but it's a very big haystack and the needle is moving

Chapter 7

Several people told me about a man who had run into the Bigfoot down by Lava Hot Springs, but it took a while to track down the story. This guy is another hunter, trapper, and woodsman who knows what he is listening to when he is out in the woods, and he is positive that his experience wasn't normal.

DEMPSEY CREEK

In the late 1970's Tom Fuller was working as a miner near Soda Springs, and being a young guy living in a smaller town, he often made a trip to the big city of Pocatello on the weekends just looking for something to do. Some of the time he would be heading back to Soda late at night and would become tired so he would pull off the road and go to sleep in his '69 Ford pickup. Fuller figured he was safe doing that because he always had a big Buck knife with him and also had his half breed Shepherd dog that was absolutely fearless.

"I ran with a wild crowd then, and I had seen that dog take down full grown men who were a little out of control," Fuller said. "On this particular day I had pulled off the road down near Dempsey Creek and tied my dog to the handle of the pickup and then went to sleep."

He said something woke him about four o'clock in the morning and it was loud and weird. "It was sure something strange," he said. "I'd never heard anything like it before, and it was moving rapidly down the mountain towards me."

He said the dog was growling and upset, so he pulled it into the cab of the pickup with him, fearing it would draw attention to him or engage whatever was out there.

"That dog could have taken on just about anything," he stated. "He was a great watchdog."

He said the noises kept coming closer down the mountain, starting out with what he describes as a musical tone, but then it changed to a higher note, moved down the scale to a scream, and then to a loud bellow.

"I thought about starting up the truck and getting out of there but those old Fords took a little persuading and they were noisy so I just sat there with the dog," he said. "He wasn't barking because he wasn't a barking dog, but he was sure growling."

The Bigfoot came down the mountain and ended up standing in the brush right in front of the pickup. Fuller said he was watching to see what was out there but he was focused at eye level. When he glanced over at the dog, it was looking UP and could obviously see something much bigger than Fuller expected. He said it felt as though the being stood there and yelled for a long time, like a standoff of sorts, but eventually the Bigfoot passed the pickup and went down the mountain.

"I don't know why it was making the noise, but it certainly woke me up," Fuller said. "I wasn't going to go back to sleep, so I sat there until daylight and then left and headed back to Soda Springs."

He said that over the course of his lifetime he has spent many, many hours outside as a hunter, trail guide, woodsman, trapper and miner all through the Rocky Mountains from Mexico all the way to Alaska, and he has never heard anything or experienced anything like that before or since.
Once he left there he never went back either. "In my opinion, it is nothing to mess with, leave them alone!" he said. "I've heard everything out in the woods and it wasn't anything native to our area."

Years later while watching a National Geographic documentary on Bigfoot he realized how amazing his experience had been. "They played a recording of the Bigfoot and it was just like what I heard only not nearly as loud. The one I heard was upset," he said. "I didn't stop or stay around or look for tracks because I didn't want to see any."

He said at the time this experience happened he didn't have a gun with him, just his big Buck knife and his dog, but ever since then he has made sure to carry a gun with him at all times. "I never go anywhere out in the woods without a firearm anymore," he said. "If someone else had been out there unarmed, it would not have turned out well for them."

This print came out so clear for being in the forest – and check out that size!

Chapter 8

It's tough to separate normal noises from strange noises that go bump in the night when you live way out in the sticks. Inna Nappo is one of those people. She lives way out from town and doesn't enjoy being outside at any time but even less since seeing the Bigfoot. This incident happened in the fall of 2017.

FERRY BUTTE

"It was in the middle of the night in the fall when I clearly heard tapping outside the front door," Nappo said. "We live way out in the country and don't have any outside porch so that was pretty creepy. All the dogs were barking and I could hear the owls in the neighborhood. The dogs would become scared and then play dead. My nephew picked one up by the head, and it just dropped like it was dead. but it wasn't."

She said they could see the Bigfoot in the trees in the backyard making all sorts of noise. Eventually, they threw rocks towards the being and it ran off.

"I hate being outside," Nappo said. "It was really, really dark that night. I used to have a bunch of dogs outside to protect me but now I am down to just two. Ever since I lost the other dogs, I have been crazy scared to go outside."

She said she has heard the Bigfoot scream in three different spots – behind the home, in front, and across the road in the field.

"There were two in the backyard near the bottoms at the same time," she said. "They were definitely not dogs, and they ran down the bottoms area still howling – a long eerie sound that is REALLY deep."

She has been hesitant to say anything about the experiences she had.

"We believe if you see them, someone will die, and if we talk about them we will see them," she said. "I don't want to see them anymore."

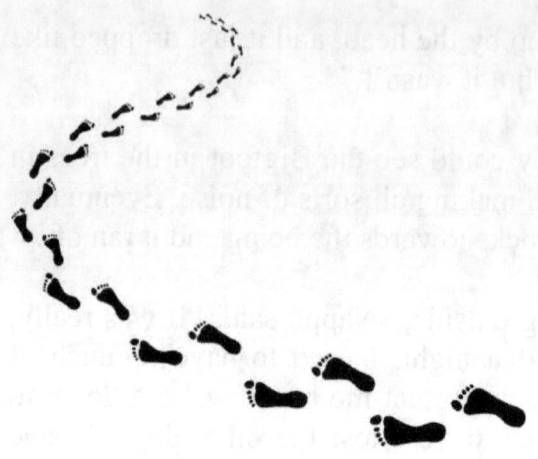

Chapter 9

Wes Douglas has a degree in wildlife biology and has worked for the US Forest Service for over ten years, but he didn't see the Bigfoot until after he left that job and started laying hardwood floors in Boise. He still loves to spend time out in the woods so periodically he plans backpacking trips out into the forest. This particular time he was on a seven-day backpacking trip in the Frank Church Wilderness near Salmon when he ran into a Bigfoot. This is that story.

FRANK CHURCH WILDERNESS

"I met Rob Bozzuto (from *Bigfoot Lives in Idaho Book One*) at a BFRO (Bigfoot Field Researchers Organization) meeting and realized there were other people who had seen the Bigfoot," Wes Douglas said. "That was a real eye-opener to me, validation of what happened to me."

Bigfoot Lives Forever…In Idaho

It was during the summer of 2017 that Wes Douglas and several friends went on a seven day back-packing trip through the Frank Church Wilderness near Salmon, Idaho. There were five of them, including one father-son team, and they were well, off the beaten path one night when they stopped at a lake without a maintained trail. There was another backpacking group on the other side of the lake, so they camped on the opposite side. It was obviously an area that wasn't used often – there were no amenities, and no indications that anyone traveled up that way.

They started setting up their camp and eventually ended up sitting around the campfire for the night. Just when they thought everything had settled down for the evening they heard a huge knock, coming from an area close by the camp.

"One of our group was a Boise policeman who wondered what we should do," Douglas said. "We finally decided not to do anything – to just pretend everything is fine. Eventually, we picked up our area and went to bed."

Douglas said he wasn't quite ready to settle in for the night so he decided to take a walk down by the lake on the side away from where they heard the knock. He intended to bring a bucket of water to extinguish the fire but as he made his way down to the water, wearing his headlamp on his head, he saw a large figure run across in front of him.

Immediately he turned around and went back to his friends and said, "I saw something, and I think some of you should go down to the water with me."

Douglas said he usually stays up later than the other hikers, but the guy in the tent next to him started snoring so it was hard to settle in this particular night.

"I wasn't asleep yet when I heard another yell right outside camp," Douglas said. "Immediately after that I heard someone (human) scream, so I jumped out of bed to check the situation out."

He said his 16-year-old son had a nightmare that night but he wasn't sure if that was what triggered the scream or was a result of the other scream he heard. Either way, no one wanted to go out and check. Later that night the Boise policeman texted him and said, "Now I believe in Bigfoot!"

Chapter 10

This story is unique in that this particular sighting wasn't just one person; it was a whole family of humans who saw the Bigfoot. They were all amazed and a little shocked. This is their story as told by one of the sisters in the group.

IDAHO CANADA BORDER

It was the summer of 2014 about seven in the evening when Amy Yorgensen saw her first Bigfoot. At the time, she did a lot of business in Canada while working for her brother's business. During that summer her brother-in-law and sister and all their kids would ride with Amy on her business trips as they drove to Banf, Alberta, Canada. This particular trip, her brother-in-law and sister were in front of her pulling a trailer with their vehicle while all the children rode with her.

She said they pulled over to sleep for the night at a campground near the Idaho Canadian border after a long day of driving, when all of a sudden, they saw Bigfoot.

"This guy walked across the road in front of the vehicle less than fifty yards away from us," she said. "He was big, huge, tall, and he walked quickly across the street. He had NO clothes on. My nieces and nephews saw him too."

She said they stopped to take a picture but couldn't get the camera out fast enough to capture a decent picture. They also spent some time looking in the woods all around the same area, but couldn't see anything – it was as though he just vanished!

"It was weird," Amy said. "I thought I must have imagined that he wasn't wearing clothes, but I saw people the same distance away and could easily tell they had clothes on."

She said her brother-in-law told them they couldn't tell anyone the story or the people would think they were crazy, but they all saw him. "It was about seven feet tall but I'm not sure if it was a male or female – I couldn't tell from that distance," she said. "It was nearly black though. It seemed like it happened so fast…"

She said when they drove into the campground where they stayed the night there were trees, but they weren't "super close" together and the ground was clear, so there was no chance to find any footprints. "We were really caught off guard, shocked actually," she said. "All the nieces and nephews slept with me that night!"

She said after that experience they spent a little time looking into Bigfoot sightings for the area near them in Alberta and found they have sightings all over that area, so what they experienced wasn't so out of the ordinary.

"At the time it was a shock," she said. "Now we can talk and joke about it; it's a fun memory!"

CHAPTER 11

I keep finding such fun stories! This one is from the Island Park area by Mack's Inn. If you have ever traveled in this area you know it is peaceful and beautiful – just what the Bigfoot and the humans like. Heidi Turner was working near Mack's Inn when she had this experience late one night.

ISLAND PARK

Heidi Turner grew up in the Island Park area so she is very knowledgeable about the roads, buildings, and mountains around there. She said the night she saw the Bigfoot she was heading home from a late night hanging out with her friends. It was 1997 in the summertime, either

Bigfoot Lives Forever…In Idaho

June or July. It was late, almost midnight, and she was driving home on autopilot on the dirt road north of Mack's Inn. She came to the T in the road and stopped; just then a Bigfoot ran across the road in front of her car.

"I was completely stopped and he just went across the road directly in front of my car, less than three feet away!" she said. "It shocked me. I sat there and locked all the doors and then slowly drove home."

She said it was such a shock to her senses that she kept asking herself "What just happened?" She said that he had to have noticed her. He had to have been aware that she was sitting there in her car – there is just no other possibility. We will likely never know why he chose to cross the road there; it might be likened to why the chicken crossed the road.
She describes him as being large and dark in color. He passed the side of the car towards the way she was coming from.
.
"It was a little chilly, and I had the windows closed, so I have no idea if there was a smell or not," she said. "I had been goofing off with my friends, but coming home I was all alone – there was no other witness."

She said she hasn't seen anything of the Bigfoot since then, but she still thinks of him each time she is on that road.

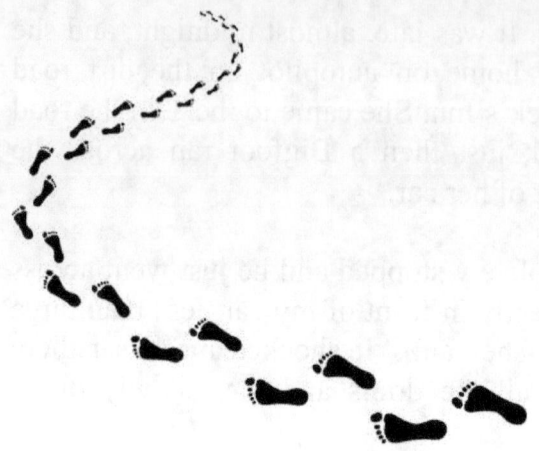

CHAPTER 12

I first met Walt Evans when he came to hear me speak, and he introduced himself. At the time he had a healthy interest in the Bigfoot and enjoyed placing trail cameras to see what he could photograph. Vans is retired and enjoyed being outside. He used his time to indulge his interest in the Bigfoot and he had been having some fun exploring areas that he thought had merit. We ran into each other at a few different events, and then he and his wife moved to a new area in Idaho where he started finding Bigfoot prints. Now it seems every spring he finds even more footprints, all of them around his home or up in the foothills near Malad.

MALAD

Bigfoot Lives Forever...In Idaho

Walt Evans was purchasing some eggs from a neighbor when something she said startled him. "Did you see the guy in your backyard yesterday?" "No," Evans said. "What did he look like?"

The neighbor said he was wearing a furry hat and a furry coat. Evans initially thought maybe someone was trying to steal something from them. He and his wife had just moved into the area four or five months previously and didn't know the neighbors well yet.

He just about forgot about the whole conversation. Then, the next day, while he was out walking around his property looking for his boundary markers, he found some footprints.

"There they were! They weren't real big, maybe thirteen or fourteen inches long," he said. "But they were barefoot, out in the slushy snow and mud – who walks around in their bare feet that time of year?"

Evans realized that he was seeing Bigfoot prints and the more he thought about it, he realized that it was likely after the neighbor's goats and chickens. Over the next few months, they had a wet spring, and he found even more footprints near his home and up in the foothills close to his home. The next year in the spring he found even more.

"I find a lot of footprints in the spring," Evans said. "There are a lot of them, but one Bigfoot could be making a lot of tracks. It was drier the next year, but I still found some more footprints up near the junipers, but they were bigger. My foot was dwarfed by it; it must have been eighteen or nineteen inches long."

He said each time he finds one footprint he expects he won't find anymore but then he finds the next set. "If you are an observant person you notice all sorts of things that are out of context – that aren't supposed to be there," he said.

That is how he has seen and photographed the Bigfoot that although he admits that most of the pictures are blurry. He said the Bigfoot seem to sit intensely still for long periods of time and watch people. When he is scanning an area, he notices something that doesn't belong there and he will pull out his scope to obtain a better look at them. Each time the Bigfoot, haven't seemed aware of him because he has been a long distance away from them.

"They imitate things – all sorts of inanimate objects like fence poles, stumps, or trees," he said. "I will set a scope on something that looks odd and will take a picture. It appears they are just sitting still, for a long time. Then when I come back, they are gone."

Evans said if he wasn't married he would just sit and watch them all day because he would seriously like to take a picture of them moving, especially one a little closer to him than farther away.

"I can see them, but they can't see me," he said. "They are just doing what they do. I don't think they can sense me when I am that far away. I'm waiting for that time when they move WHEN I am watching."

He said he has come to the conclusion there are more of them out there than most people realize, but they are also migratory, at least that is what he thinks.
"I think they go where the food goes," he said. "They follow the deer."

Evans said he has spent a lot of time thinking about what he has seen; his knowledge has come after hours of observation of them just doing what they normally do. "It's just a shot in the dark really," he said. "I'm just guessing from the tracks I have seen in the spring when there are a lot of them. I think people would be surprised how many there actually are."

He said the Bigfoot watch people all the time and they are smart, so they must realize there are fewer people around at night and choose to use that time to move around.

"They are smart, and they can sit still for a extraordinarily long time," Evans said. "Besides, they have the whole forest to hide in; it's no wonder people don't see them all that often."

CHAPTER 13

Mount Putnum has long been a place of unusual Bigfoot activity, but it is also protected by members of the Shoshone Bannock Indian tribe. Members of the tribe have often told tales of the Bigfoot, but Adrian Jody Edmo was able to spend nearly three months in close proximity to a Bigfoot family, along with other members of his own family.

It's an amazing story that some would envy, but others won't even believe. We have a lot to learn from each other, especially from those of us who see and experience adventures with the Bigfoot that others don't.

MOUNT PUTNUM

Adrian Jody Edmo took his wife to camp near Mount Putnam a few years ago and had what some would think was an enviable experience – Edmo stayed near a family of Bigfoot. There was an older male and female and three juveniles, and he was able to see them and watch them for half-an-hour at a time.

"We were up by Mount Putnum near Five Points for nearly three months in the fall," said Edmo. "We ran into an entire Bigfoot family. We learned to tell them apart and which one was which when we recognized their footprints."

There was one Bigfoot making a twenty-two-inch print, an eighteen-inch print, a smaller one with an eight-inch footprint and two others whose footprints often were jumbled together. The Edmo family saw the Bigfoot family regularly and would often interact with them and leave them gifts.

"We would leave them deer and elk meat and occasionally potatoes, fish, or wild berries," Edmo said. "We would also play music for them."
He said they were all brown and black like an elk, but with a silvery sheen to their hair, especially when it was wet. "It was rather pretty!"

Edmo called the biggest one "Old Man" as he was obviously older than the others, and one day the old man came all the way up to where Edmo had parked his pickup near their camp. He said they would go up and camp in the same area each time they went to the mountains. They would usually stay until they became uncomfortable, ran out of supplies, or felt the need to leave, and then they would move their camp downhill.

"They would follow us down the hill," he said. "They startled the horses, but they would come in and take the meat and corn I would leave for them."

He said when his grandsons and nephews would come with him to hunt; they would always leave the Bigfoot the heart, liver and kidneys of their kill.

While this experience seems fanciful, Edmo said the Bigfoot have been up there as long as he has been going up there – all of his 72 years.

"They've been there since I was a kid, they don't scare me. I would ride all of the old horse trails all over," Edmo said. "They come by us at our feedlot (on the reservation) in the spring and the fall as they are traveling back to Mount Putnam across the reservation. A lot of people see them, but we don't bother them."

He said while the Bigfoot he has seen usually have brown eyes, the ones with the red eyes can't be trusted.
"They are more aggressive," he said. "They eat more meat than the others. The peaceful ones usually have brown eyes."

Edmo said he has always said prayers to them as the Bigfoot have their own medicine, but that he has also seen them disappear right in front of people.

"We just go about our business," he said.

Bigfoot saw me, but no one believed him because the pictures came out blurry…

Chapter 14

Ron Harper told me his brother David Harper and also a friend, Craig Kunz, have had experiences with the Bigfoot in the same area near Bear Lake where they all grew up as children.

That happens sometimes, especially back in the days when families spent more time outside than they did in the house. Ron told me about Craig's experience with the Bigfoot, so I tracked him down to find out what he saw because it happened at a different time and place than the Harpers experience.

Kunz was cutting firewood in the canyon between Ovid and Preston, Idaho. This particular story happened when Craig and his cousin were out in the woods together.

OVID

Bern, Idaho is a little community, not bigger than a bend in the road, and it is just down the road from another small community called Ovid. Most people don't even notice the little town as they travel the road north of Bear Lake unless they are hunting or camping. The children who grow up there tend to spend their time in the woods - sometimes being in trouble, but usually making their own fun. Such was the case with Craig Kunz and his cousin. They were camping in the canyon between Ovid and Preston, Idaho and Kunz said they had been cutting firewood for the year.

"We were on the hill, cutting firewood, when our dog just went crazy – barking and growling. He was a little yapper of a dog, but we looked and we couldn't see anything," Kunz said. "It was just weird because he wasn't known to start barking without a good reason."

The next night, he and his cousin went for a hike over the ridge from their base camp and away from safety.

"We both saw something big and dark in the brush," he said. "It scared us – but we were more uneasy than anything else. It was like being extremely fearful of something unknown or unseen I just don't understand but the hair stands up on your neck."

Kunz shot his BB gun into the bush and they ran like crazy back to camp. They didn't tell anyone about it until years later and then many people thought they just made it up.

"It was a weird experience," he said. "But I was only eleven and didn't have anything to compare it to. Normally I wouldn't shoot a gun at something that I couldn't see. But the whole experience was kind of unnerving. We didn't know what was out there but it made us uptight and nervous."

Chapter 15

I had heard about Dave Erlanson several times before I had the opportunity to meet him as his reputation had preceded him. He is an amazing hunter and feels more at home in the woods than most people. This in spite of the fact that he has a master's degree in education and at one time was a successful schoolteacher. He grew up in Pennsylvania and then moved west to Idaho where he lives in a cabin nestled way out in the woods – right out where he loves to be.

PALISADES

Dave Erlanson has been outside hunting or trapping since he was a young child. He's a professional guide with many years in the backwoods under his belt now and several impressive records for his bear kills. He didn't

Bigfoot Lives Forever…In Idaho

believe in Bigfoot, but he'd heard the stories and laughed.

That all changed a few years ago on a hunting trip up above Palisades Dam.

"It was 6:30 in the morning, when this thing screams near me," he said. "Then the scream changed to a roar and then a growl. It was so loud and felt so close that I jumped three to four feet in the air. I'm only five feet tall so that was quite a feat!"

He said he felt the reverberation of that sound through the entire fiber of his being – it felt that it rattled his chest down to his shoes. For the first time in his life he pulled his gun off his shoulder and took the safety off without a target to shoot.
"I've heard elk fight and grizzly fight. I've heard all sorts of wild cats from all over the world but I'd never heard anything like that before," he said. "It reverberated through my chest like the feeling when you are standing too close to a big speaker."

Then something funny or odd happened. "I was standing in a clearing and it was early enough that the timber was dark around me but the meadow in front of me wasn't. As soon as that Bigfoot roared, everything broke lose – every coyote lit up and started howling and a whole lot of deer came running past me," he said. "It was right then I climbed a tree and figured I would just wait it out."

He said he started to believe that maybe there might be some truth to the stories about Bigfoot he has heard.

"I'd stake my life on what I heard. I have had a grizzly tracking me but I haven't been that scared before."

Shortly afterward he was staying in a campground in Elk City and got up to use the bathroom. "I heard a crack that came from behind the tent," Erlanson said. "The next morning I found big footprints behind the tent less than ten feet away."

This year in May he found further proof that there was something else in the woods. "I had been up Bear Creek with a friend who was setting out bear bait, and we were on our way back down the mountain when I saw something that looked funny. I thought it might be a grizzly track," Erlanson said. "He stopped the truck so we could look around, and there were some perfect Bigfoot footprints."

They measured the prints and they were eighteen inches long. The stride where the Bigfoot had jumped across the road was 87 inches from the heel of one foot to the toes of the other. He was able to make a casting of the footprint and it is impressive. He said they found a dead elk down below the tracks and most of the prints were grouped where there was a good vantage point overlooking the road.

Bigfoot Lives Forever…In Idaho

The next morning he took another friend out to show him the prints, and they found a set of two more prints – a big set and a smaller set beside it.

"They were deeply sunk in the mud and were about twelve to fifteen inches long. It had rained the day before, just after I had found the other tracks," he said. "But these tracks were about four and a half miles away in a straight line from the other ones."

Erlanson has been a hunting guide in Idaho over thirty years and has also led hunters into the wilds of Alaska, Wyoming, Utah, and Pennsylvania. He holds the record for "most black bear kills in Pennsylvania," having brought down 50 of them. Even with all of his experiences, running into a Bigfoot is still one of the most unique experiences to have happened to him.

"I didn't believe in Bigfoot; I thought it was a hoax. Now I believe it and I look at every bear track differently," he said. "I haven't seen one in person yet but I will one of these days. I know they are up there. It's one thing to think you see it, and it's another thing entirely to have one scare you half to death! Why lie when the truth is so much more amazing?"

Bigfoot print found in the Sawtooth Mountains. It isn't particularly big as the woman's shoe is a size 8, but it still looks impressive.

Chapter 16

Occasionally I find stories in unlikely places, right under my nose so to speak. That is what happened with this story. Bea Cowger lives in the town of Pocatello but as the town has grown her home has gradually been associated more with the outskirts than the actual town. Still, it is unusual to find a Bigfoot story so close to a congested area but that is what happened here.

POCATELLO

A few years ago Bea Cowger's husband built a fence along the ditch bank near their home and then planted some lilies along the fence. One day he found those lilies all knocked over, and it looked like someone just got upset with them and destroyed the area he planted them in.

Bigfoot Lives Forever…In Idaho

"Years later when we realized that the Bigfoot were in the area, we figured out that they must have been a little upset, finding the fence and then the flowers right on their pathway," she said. "But they also ate all of our peaches along the ditch bank."

They had planted the peach tree and were anxiously watching and waiting for the day when they would have fresh peaches – they checked them every day in anticipation. Then one day they were all gone – even the hard ones, and there was nothing on the ground. Nothing.

Later in the year when school was ready to start, they had a chance to get a clear look at the Bigfoot. That particular morning Bea's husband was in the kitchen making coffee, and she was just waking up. They both clearly saw someone they thought was a child walking to school, but it was early to be headed to school, plus there was no way to get through on the ditch bank to go to school since the fence had been built. Besides, they had parked their camper trailer next to the fence making it doubly hard to navigate that space. Apparently the Bigfoot didn't know that as he first tried to go on the ditch bank, and it seemed to be blocked. So he walked around the camper and the Cowgers both saw him clearly standing near the door of the camper.

"I figure he must have been about eight and a half feet tall because his head and shoulders appeared clearly above the door," she said. "It seemed to be surprised to find the way blocked, and that is why it detoured all the way around the camper to pass that area and continue down the ditch."

The Cowgers had a previous experience with the Bigfoot prior to their experience at their home in Pocatello, but it actually happened when they were traveling through Yellowstone Park.

"We were about halfway to the park when I realized I was getting the flu," Bea Cowger said. "I was pretty miserable, so when we reached the park we pulled over in a parking lot under some trees and pulled out the bed so that I could go to sleep and try to feel better."

She was awakened later by the sound of someone scraping their fingernails across the top of their camp trailer – all the way across in a continuous motion. And there wasn't any wind to move the tree branches.

"It scared me so much that we moved the trailer down to another spot where there weren't any trees or bushes," She said. "Even in that original spot there weren't any branches near enough to make the noise we were hearing. We never looked to see if there were scratch marks on the top, but the only explanation I could think of was that something with claws had been running its nails across the top of the camper. It made for an interesting vacation."

They look like a cross between a Native American and a troll with a little bit of logger thrown in for good measure
-description of Bigfoot

Chapter 17

I met Luke and his dad when they came to a Bigfoot event in southern Idaho. I had met his father before so I talked to him for a moment and asked what their interest was in the Bigfoot. He mentioned that his son Luke had seen the Bigfoot while they were out in the woods the year before, so later I tracked down Luke to tell me the rest of the story.

SAWTOOTH MOUNTAINS

It was August in 2015 when Luke Pendlebury and his brothers convinced their father to take them out hunting for something bigger than bunnies. They headed out past Howe up one of the canyons and eventually found a good place to set up camp up in the Sawtooth Mountain range. Leaving their dad in camp to set up the tent, the three boys took the two-seater UTV with Luke in the bed of it to go scout around. The youngest brother, who has surprisingly great eyesight for hunting, saw something and yelled, "Stop! There's one right there! Stop, stop, stop!"

The deer started running and he took a shot at it and missed. He took another one and missed again so Luke got out and took a shot at it and hit it. While one brother took the UTV back to camp to collect their father, the other two brothers hiked up to the carcass to wait. They finally came back and loaded the carcass to take back to camp.

They hung it and washed the blood out and decided that they were going to get the youngest brother another deer at least. They drove around some more and found a trail that brought them uphill.

"Coming up to the left of the trail you could see a 1960's Jeep truck wedged between some trees. It must have been salvaged by someone," Luke said. "Further down the trail we came out to this beautiful clearing, like something out of National Geographic magazine. In the top of the clearing there was an old cabin like one a trapper might have built out of logs that were about a foot wide."

He said they just marveled at how beautiful the clearing was, and eventually they started looking for tracks and found some elk tracks leading down to a stream in the little clearing. As they sat in the UTV deciding where to go next, a hunter they had never seen before approached them and he was angry. He told them they were in there illegally and that they shouldn't be driving their UTV around that area.

"We didn't really trust this guy because he was what we would call a "Cabella's hunter", where he looked like he had just come out of a Cabella's store with completely new gear," Luke said.

They tried to plead their case, but the man just became angrier. They thought he was going to start shooting at them. Just then, they happened to look up the hill and they saw what appeared to be a large man, walking across the hill line they had just left. They asked the angry hunter if that was his man and he said, 'Ya, and he is aimed right at you," but the being didn't have a gun.

Luke said it looked like a black silhouette of a figure that moved almost exactly like the Bigfoot in the Patterson Gimlin film. "It was about as tall as one of the trees he walked across, and we estimated the tree to be about eight or nine feet tall," Luke said. "It was strange because when we went to look for this figure, it seemed there wasn't a trace. Only the memory remains."

Luke said that his dad and brothers also saw the Bigfoot, but the hunter didn't turn around to look, either because he actually did know it was there or because he just wanted to be belligerent. Either way, they chose to ignore him and disengage.

They backed away and left the area without further conversation. They only stayed up in the mountains a night or two, and they didn't see either the hunter or the Bigfoot again.

Luke said he had heard that area was a hotspot for Bigfoot sightings so he is happy to add his story to the mix.

I'm no longer a believer, I'm a knower!

-David

CHAPTER 18

Terry Baddley loves spending time outside – it's his favorite place to be when he isn't working. He has had a couple of fun experiences with the Bigfoot while outdoors both in Idaho and in Utah throughout the years, and they all serve just to let him know there is something out there bigger than all of us.

STRAWBERRY SPRINGS

"We were out camping in the area between Preston and Montpelier by Strawberry Springs," Baddley said. "We had camped for the night and my brother was under the truck sound asleep. My friend and I were sleeping in the back of the truck. In the middle of the night the dogs started growling, and my brother told them to 'go get it!' so the two dogs took off."

When the dogs caught up with whatever it was, they heard the most ungodly noise – a cry, a scream, and then a yelp. Terry said that he got the guns out of the truck, and when the dogs came back they just started rolling on the ground, acting just terrified.

The Bigfoot followed the dogs in for about thirty yards, and he yelled at it "We've got guns and we will shoot!" It kept coming so Baddley fired into the brush.

"I don't know how smart that was," he said, "but after a while it turned and left us alone."

MANTUA

In the summer of 1977 Baddley was dating his first wife, and they were driving around outside of Mantua, Utah to the east of Brigham City.

"It was after midnight, and we drove the backroads up a dirt road under a hillside of scrub brush," he said. "We were parked, looking outside the window. Through the moonlight I could see the black figure of a man walking up the hill."

He said he thought to himself 'Someone's coming!' and figured they would walk up by the truck and give him a hard time, but no one came.

"I had a negative feeling and told my girlfriend to hold on. I turned right around and came down the hill, half expecting to see a vehicle, but we never saw one," he said. "There were no hiking trails, and there was nowhere for anyone to walk from. I wasn't thinking about it being a Bigfoot or anything like that then, but I should have been able to tell if it had clothing on. It was just strange."

He said about twenty years later he read a BFRO account that happened just a little way from there – within just a few miles, and maybe only a mile as the crow flies. "I wished I would have had the presence of mind to gather more details," he said. "But I assumed it was a man. I won't say it was a man or a Bigfoot…"

Years later a big group of friends had gone into the Uintah Mountains and Baddley heard the tree knocking sounds. "People say that is what they sound like and if so there were over 40 Bigfoot outside the camp," he said. "They weren't in the same place, but there were several beings all knocking at once."

Chapter 19

It is a given that Bigfoot is likely one of the biggest beings you will run into in the woods, but he seems a whole lot bigger when you are a little kid. Scott Smith had an interesting experience as a youngster at the Willow Creek Hot Springs camp-ground near Featherville in the Sawtooth National Forest.

WILLOW CREEK

Each year the Smith family goes camping in the same area, and they have for many years. It is a sweet little camping ground at the Willow Creek Hot Springs near Featherville in central Idaho. Several years ago when Scott Smith was a young man of four, the Scott family were on the annual family camping trip when he ran into his first Bigfoot.

"It was late at night and I needed to use the bathroom," Smith said. "I went down the path to the little one-holer porta potty latrine and did my business."

He said he opened the door to the latrine and started walking back to the family camp when he saw a huge Bigfoot, walking down the road towards him. He didn't know that was what it was though, not until a couple of years later. He just knew that he ran into a very tall person who looked like a scary man, wearing a big fur coat.

"I immediately turned around and ran back into the latrine and closed the door," Smith laughs. "I remember thinking that the man was pretty scary and that he wore a fur suit. I also wondered why anyone would wear fur pants…" Although it looked like he was wearing something furry, Smith said that it looked glittery in the moonlight. He didn't smell anything different, but then he admits he has a terrible sense of smell since a childhood accident.

He said it took a few years but when his brother was talking about Bigfoot one day it all clicked. He finally realized what he had seen that late night on a camping trip.

"We spend a lot of time out there each year, and I love hiking the hills around there. I have never seen anything like that before or since," Smith said. "I have also asked a lot of people about the Bigfoot, and they have never mentioned anything similar happening."

He keeps his eyes open when he is up there each summer as he hikes, but it still sticks in his mind, perhaps because it was so surprising.

"It was bigger than anything I have ever seen – really tall and furry from head to toe!" he said. "It was just so surprising – I didn't expect to run into anything on a trip to the bathroom."

ABOUT THE AUTHOR

Becky Cook is an Idaho native who loves to write. She has written for numerous newspapers and magazines throughout the United States and has reported local, regional, and national news. She has been fascinated with Bigfoot sightings and stories all her life and it is a favorite subject for her eight children as well. The Bigfoot stories have delighted many of her friends and family for many years, and now they are being retold here for you – her readers.

This book is the third of the *Bigfoot Lives in Idaho* series. To purchase her books go to www.bigfootlives.com. If you have had a sighting and would like to share that information, please write to her at: becky@beckycookonline.com

ABOUT THE ARTIST

Brandon Tennant was born in Miles City, Montana and raised in West Yellowstone, Idaho. He has been drawing all his life. He became interested in Bigfoot when he was a child and first heard of a sighting in his area. He drew the picture used on the cover of this book and is well known for his Bigfoot likenesses, which he puts on T shirts available online and at his business at 348 N 3rd Ave in Pocatello, Idaho, where he runs Falling

Bigfoot Lives Forever…In Idaho

Rock Productions. He also runs www.Sasquatchprints.com.

www.ingramcontent.com/pod-product-compliance
Lightning Source LLC
Chambersburg PA
CBHW070810220526
45466CB00002B/625